FUN FACT FILE: HABITATS

# 20 FUN FACTS ABOUT FOREST HABITATS

BY CHAR LIGHT

Gareth Stevens PUBLISHING

**Please visit our website, www.garethstevens.com. For a free color catalog of all our high-quality books, call toll free 1-800-542-2595 or fax 1-877-542-2596.**

Library of Congress Cataloging-in-Publication Data

Names: Light, Char, author.
Title: 20 fun facts about forest habitats / Char Light.
Other titles: Twenty fun facts about forest habitats
Description: New York : Gareth Stevens Publishing, 2022. | Series: Fun fact File: Habitats | Includes index.
Identifiers: LCCN 2020036348 | ISBN 9781538264430 (library binding) | ISBN 9781538264416 (paperback) | ISBN 9781538264423 (6 pack) | ISBN 9781538264447 (ebook)
Subjects: LCSH: Forest ecology–Juvenile literature.
Classification: LCC QH541.5.F6 L55 2022 | DDC 577.3–dc23
LC record available at https://lccn.loc.gov/2020036348

First Edition

Published in 2022 by
**Gareth Stevens Publishing**
29 East 21st Street
New York, NY 10010



Designer: Michael Flynn
Editor: Kate Mikoley

Photo credits: Cover, p. 1 (main) Miguel Navarro/DigitalVision/Getty Images; file folder used throughout David Smart/Shutterstock.com; binder clip used throughout luckyraccoon/Shutterstock.com; wood grain background used throughout ARENA Creative/Shutterstock.com; p. 5 Zocha_K/iStock/Getty Images; p. 6 Rocky89/E+/Getty Images; p. 7 (pine tree) DNY59/E+/Getty Images; p. 7 (pine branch) undefined undefined/iStock/Getty Images; p. 7 (pine cone) sbk_20d pictures/Moment/Getty Images; p. 7 (maple tree) Tetra Images/Getty Images; p. 7 (maple leaf) Grzegorz Wozniak/EyeEm/Getty Images; p. 7 (maple seed) Daniel Haug/Moment/Getty Images; p. 8 Ann Cutting/The Image Bank/Getty Images; p. 9 Darrell Gulin/The Image Bank/Getty Images; p. 10 Misha Kaminsky/E+/Getty Images; p. 12 Manfred Gottschalk/Stone/Getty Images; p. 13 VCG/Visual China Group/Getty Images; p. 14 odmeyer/iStock/Getty Images; p. 15 Chadked/iStock/Getty Images; p. 16 Jami Tarris/Stone/Getty Images; p. 17 Anjan Upadhya/500Px Plus/Getty Images; p. 18 Robert Winkler/iStock/Getty Images; p. 19 Sheryl Watson/iStock/Getty Images; p. 20 KenCanning/E+/Getty Images; p. 21 Stickney Design/Moment Open/Getty Images; p. 22 simon's photo/Moment/Getty Images; p. 23 Corey Hardcastle/500px/Getty Images; p. 24 Jillian Cooper/iStock/Getty Images; p. 25 Denis Debadier/Photononstop/Getty Images; p. 26 jean1979/Shutterstock.com; p. 27 fhm/Moment/Getty Images; p. 29 Marius Roman/Moment/Getty Images.



Printed in the United States of America



CPSIA compliance information: Batch #CSGS22: For further information contact Gareth Stevens, New York, New York at 1-800-542-2595.

# CONTENTS



Words in the glossary appear in **bold** type the first time they are used in the text.

# LET'S LEARN ABOUT FORESTS!

Our planet is made up of six types of biomes. These are places on Earth where certain living things, well, live! These six major kinds are freshwater, marine, desert, grassland, tundra, and forest biomes.

Forest biomes are one of the biggest biomes. One-third of Earth's land is covered in forests! Forest biomes are made up of trees and other woody plants, such as bamboo. Of course, you can find lots of other living things in forest biomes too. There are bugs, **mammals**, birds, **fungi**, and sometimes fish!

Bamboo forests are home—and a food source—to many animals. Red pandas are raccoon-like animals that eat mainly bamboo. Bamboo is a tall, thick grass.

# I'M ON LEAVES AND NEEDLES!

FUN FACT: 1

## SCIENTISTS BREAK TREES INTO TWO MAIN GROUPS—THE ONES WITH LEAVES AND THE ONES WITH NEEDLES.

Deciduous trees have flat leaves that they lose in the fall. Coniferous trees have needles. They keep their needles year-round.

Maple trees are deciduous. In fall, their leaves change from green to shades of yellow, orange, and red.

# DIFFERENCES BETWEEN DECIDUOUS AND CONIFEROUS TREES

These are the main ways you can tell deciduous trees and coniferous trees apart.

## FUN FACT: 2

## PINE CONES CAN STAY ON A TREE FOR UP TO 10 YEARS!

The cones are covered in scales. The scales stay closed when it's cold and wet. They open when it's warm and hot. Then the seeds can spread.

Deciduous trees often do well in warm, wet **climates**. Coniferous trees tend to live in cold climates. In many places with four seasons, both kinds of trees can grow.

FUN FACT: 3

## FORESTS WITH BOTH DECIDUOUS AND CONIFEROUS TREES MAY BE HEALTHIER THAN FORESTS WITH ONLY ONE KIND.

Biodiversity is a big word for "lots of different species living together." Forests with lots of different tree species, or kinds, grow faster. They're also safer from bugs and sickness.

# THE THREE FORESTS

FUN FACT: 4

## THERE ARE THREE MAIN KINDS OF FOREST HABITATS—TROPICAL, TEMPERATE, AND BOREAL.

Forest habitats are grouped by climate. Tropical forests are warm and wet. Temperate forests have seasons and get colder in winter. Boreal forests have the longest, coldest winters.

Boreal forests are also called "taigas." Most trees in these forests are coniferous.

# FOREST HABITATS AROUND THE WORLD

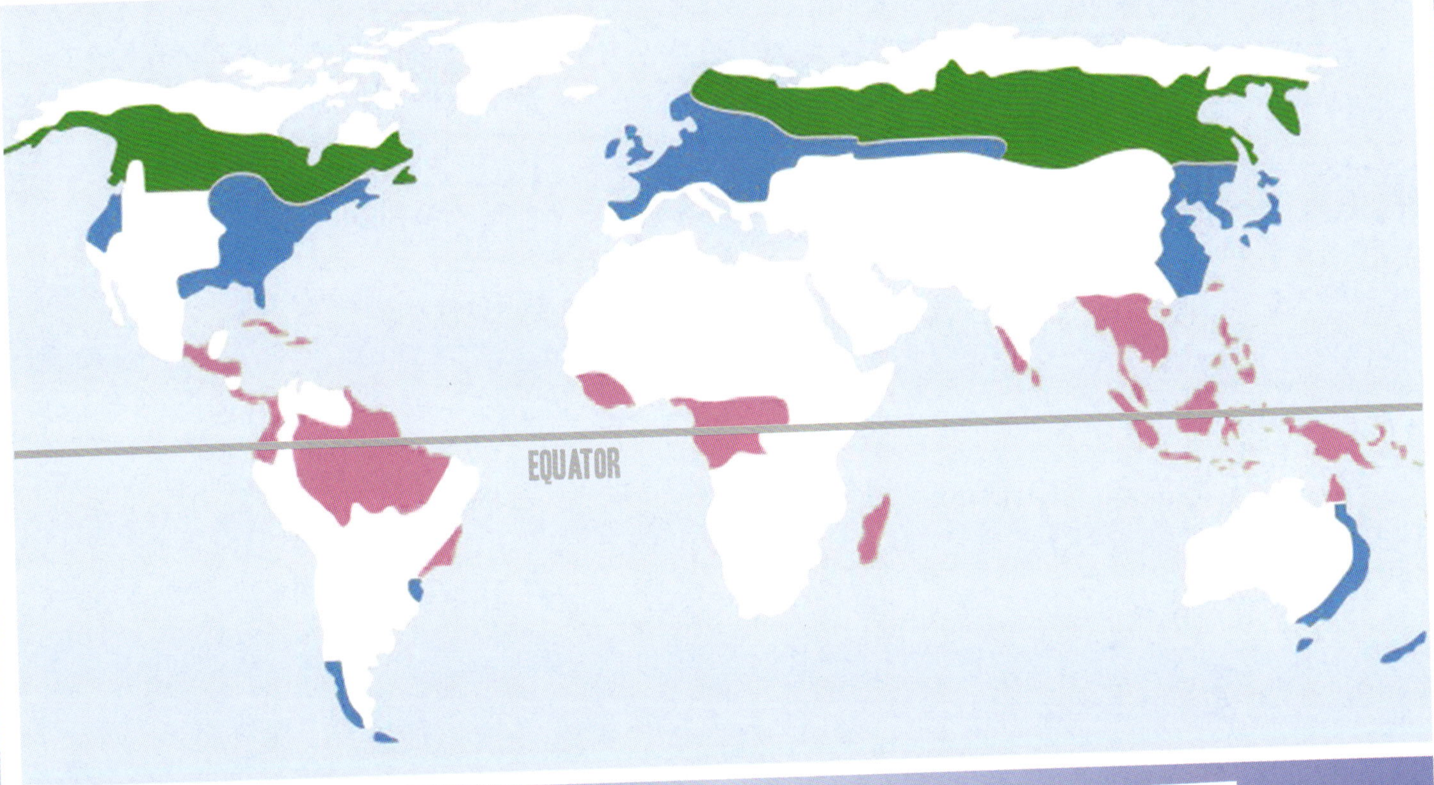


**BOREAL FOREST**
(BETWEEN LATITUDES 50° AND 60° NORTH)

**TEMPERATE FOREST**
(BETWEEN LATITUDES 25° AND LATITUDES 50° NORTH AND SOUTH)

**TROPICAL FOREST**
(BETWEEN LATITUDE 23.5° NORTH AND LATITUDE 23.5° SOUTH)

Earth's climate is warmest and wettest at the **equator.** That's why tropical forests grow close to there! The farther away you get from the equator, the colder and drier the climate becomes.

# LET'S GET TROPICAL!

## FUN FACT: 5

### IN AN AREA OF JUST 0.4 SQUARE MILES (1 SQ KM), A TROPICAL FOREST CAN HAVE OVER 100 DIFFERENT SPECIES OF TREES!

Most temperate forests have only three to four tree species in the same amount of space.

Many lianas, or vines, grow in tropical forests.

The Chinese pangolin is an **endangered** animal. It lives in tropical forests in Asia. Its back is covered in scales for protection, but its belly is soft.

FUN FACT: 6

## CONIFEROUS TREES DON'T JUST GROW IN COLD CLIMATES— THERE ARE TROPICAL EVERGREEN FORESTS TOO!

The dry evergreen forests of Indochina are home to many big animals. Asian elephants, tigers, Malayan sun bears, Javan rhinoceroses, and river dolphins all live here.

# IT'S ALWAYS RAINY IN AMAZONIA

## FUN FACT: 7

### MORE THAN HALF OF ALL PLANTS AND ANIMALS ON EARTH LIVE IN TROPICAL RAIN FORESTS!

There are over 40,000 species of plants in the Amazon rain forest alone. This rain forest, sometimes called Amazonia, is the biggest tropical rain forest on earth.

Different areas in the Amazon receive between 60 and 140 inches (152 and 356 cm) of rainfall each year.

The rain forest floor is covered in dead plants and leaves. Fungi grow on the floor and break down the dead plants. Many bugs live in this **decay.**

FUN FACT: 8

## ONLY 2 PERCENT OF SUNLIGHT REACHES THE TROPICAL RAIN FOREST FLOOR!

Tropical rain forests have thick **canopies** that block most sunlight. Trees that reach above the canopy make up the emergent layer. Plants growing below the canopy and above the floor make up the understory.

# RULED BY WATER

## FUN FACT: 9

**MONSOON FORESTS GET BETWEEN 100 AND 200 INCHES (2.5 TO 5 M) OF RAIN PER YEAR!**

Monsoons are seasonal changes in wind direction along coastlines that result in changes in rainfall. Forests that grow in areas with monsoons have a long, dry winter season and a wet summer season.

This funny-looking fellow lives in monsoon forests. It's a proboscis monkey! Its big nose makes it attractive to the other monkeys.

Many kinds of trees couldn't survive in such salty waters. Mangrove roots filter out, or remove, much of the salt before the water enters the tree.

FUN FACT: 10

## MANGROVE FORESTS SEEM TO STAND ON WATER!

Mangrove trees and shrubs sit on a thick tangle of roots. Their roots stick out from the water. Mangroves grow in warm, salty water along coasts.

# DON'T LOSE YOUR TEMPER-ATE FOREST!

**FUN FACT: 11**

**IN TEMPERATE DECIDUOUS FORESTS, SOME INSECTS CAN'T LIVE THROUGH THE WINTER—BUT THEIR EGGS CAN!**

Some forests, like temperate forests, have winters that can get below freezing. Insects that can't take the cold will lay eggs before winter, then die. The babies are born in spring.

The canopy in a temperate deciduous forest is much thinner than a rain forest canopy. This means more light reaches the forest floor. Lots of different plants and animals live here.

The takahe is a rare bird that can't fly. It mostly lives in New Zealand's grasslands but at times moves to the forest, where it's easier to find food and shelter during heavy snowfall.

FUN FACT: 12

## BATS ARE THE ONLY MAMMALS NATIVE TO NEW ZEALAND'S TEMPERATE FORESTS.

The long-tailed bat and short-tailed bat are New Zealand's only native land mammals. However, Europeans brought cats, rats, possums, and other land mammals to the area. These **introduced species** hurt some native birds and reptiles.

# INTO THE RAIN FOREST

## FUN FACT: 13

### "SPIRIT BEARS" ARE FOUND ONLY IN THE WORLD'S LARGEST INTACT, OR UNBROKEN, TEMPERATE RAIN FOREST.

Kermode bears, often called spirit bears, are a rare type of bear. They have white fur! They only live in the Great Bear Rain Forest in British Columbia, Canada.

Spirit bears get their pale fur from a rare **gene** they get from their parents. People who study animals think there are fewer than 400 spirit bears left in the world today.

Redwoods are so big that they become habitats for other living things! Bugs, birds, and other animals live in redwoods. Other plants grow on or near redwoods too.

FUN FACT: 14

## THE TALLEST TREE IN THE WORLD IS LOCATED IN THE TEMPERATE RAIN FORESTS OF CALIFORNIA.

This coast redwood tree is 380 feet (115.8 m) tall. Redwoods only live in the temperate rain forests of California and Oregon.

# BRILLIANT BOREAL

FUN FACT: 15

## THE BIGGEST FOREST HABITAT ON EARTH IS CANADA'S BOREAL FOREST.

This huge forest takes up 1.2 billion acres (485 million ha). The land it covers is 25 percent of all intact forest land on Earth. It's even bigger than the Amazon rain forest!

You can sometimes see the aurora borealis over boreal forests. "Boreal" means northern. That's why the aurora borealis is also called the northern lights!

Many freshwater fish make their homes in the lakes and rivers of boreal forests.

FUN FACT: 16

**THERE IS MORE FRESH WATER ABOVE THE GROUND IN CANADA'S BOREAL FOREST THAN IN ANY OTHER LOCATION ON EARTH!**

Rivers, lakes, and wetlands are all part of Canada's boreal forest habitat. Wetlands are covered in standing water at least part of the year.

# CARIBOU, TIGERS, AND BOARS—OH MY!

## FUN FACT: 17

## CARIBOU ARE A KEYSTONE SPECIES OF CANADA'S BOREAL FORESTS—AND THEY'RE IN DANGER OF DYING OUT.

A keystone species is one that a habitat needs to stay healthy. When a keystone species is in trouble, the whole habitat is in trouble.

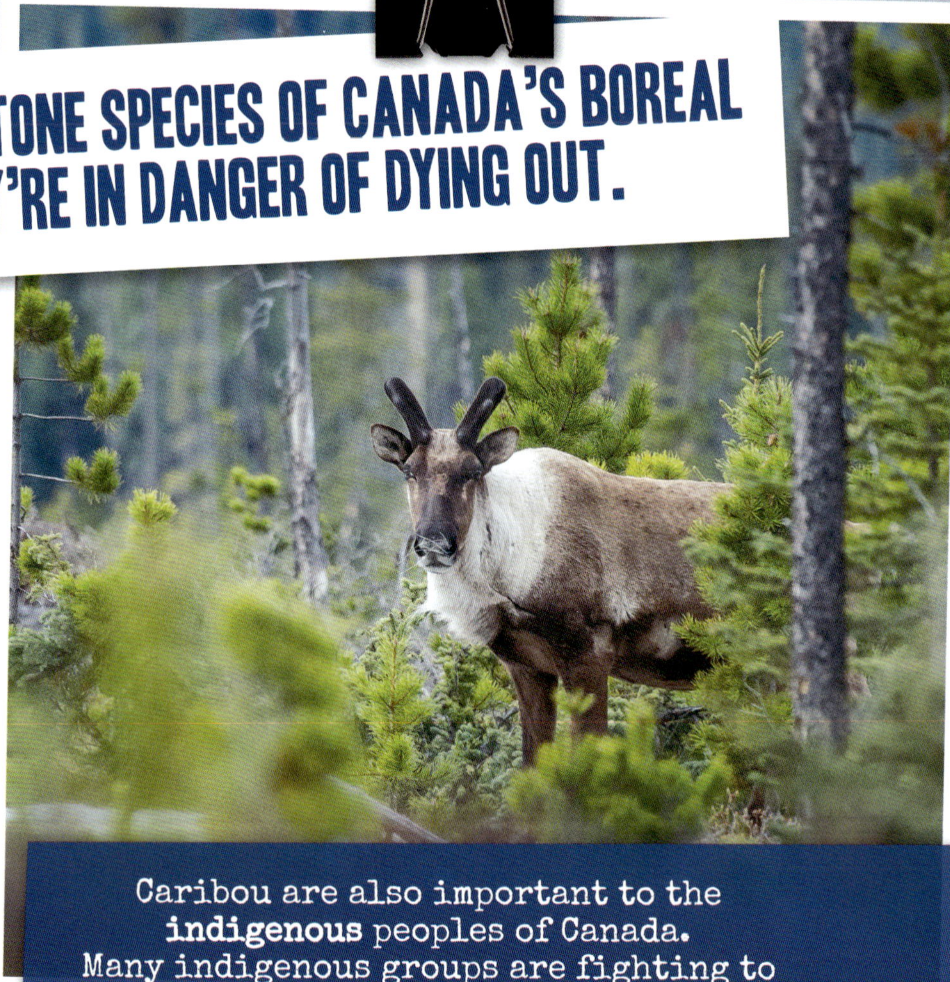

Caribou are also important to the **indigenous** peoples of Canada. Many indigenous groups are fighting to protect caribou and Canada's boreal forest.

It's illegal to hunt Siberian tigers, but people still do it. Poachers are people who hunt and kill animals illegally.

FUN FACT: 18

## SIBERIAN TIGERS LIVE IN THE BOREAL FORESTS OF RUSSIA.

There are likely fewer than 500 Siberian tigers in the wild. Tigers eat big animals like deer and boars, which are a kind of wild pig. Humans are hunting food that Siberian tigers need.

# BIZARRE BIOMES FOR THE BRAVE

## FUN FACT: 19

## IN SOME PLACES, YOU CAN FIND FORESTS FULL OF THORNS!

Thorn forests are found in North America, Africa, South America, and Australia. These forests are made up of mainly short, thorny trees. These commonly dry areas have a short rainy season when some plants flower.

Thorn forests in South America are often called caatinga. It might not look like it, but over 1,200 plant species grow in Brazil's caatinga.

The J-shaped trees in the Crooked Forest may look weird, but they're thought to be healthy. They can reach around 50 feet (15.2 m) in height.

FUN FACT: 20

## THERE'S A FOREST IN POLAND WITH J-SHAPED TREES!

Around 400 strange pine trees grow in the Crooked Forest. No one knows how they got their odd shape. One guess is that there was a heavy snowfall that bent the trees when they were young.

# MAKE LIKE A MAQUIS!

Mediterranean forests have hot, dry summers and can sometimes look more like deserts than forests. They're mostly made up of coniferous shrubs and are found in places around the world with climates similar to that of the area around the Mediterranean Sea. When they grow near the Mediterranean Sea, Mediterranean forests are often called maquis.

If you liked learning about forests, don't stop exploring! Make like a maquis and brush up on your knowledge of Earth's other biomes.

Maquis are a kind of scrubland. These are environments with lots of brush, or plants like bushes and shrubs.

# GLOSSARY

**canopy:** the upper branches of a forest

**climate:** the average weather conditions of a place over a period of time

**decay:** rot

**endangered:** in danger of dying out

**equator:** an imaginary line around Earth that is the same distance from the North and South Poles

**fungus:** a living thing that is somewhat like a plant, but doesn't make its own food, have leaves, or have a green color. Fungi include molds and mushrooms.

**gene:** a tiny part of a cell that is passed along from one living thing to its offspring and decides specific features in the living things

**habitat:** the place or type of place where a plant or animal naturally or normally lives or grows

**indigenous:** native

**introduced species:** species humans bring to an environment where the species are not native

**mammal:** a warm-blooded animal that has a backbone and hair, breathes air, and feeds milk to its young